FUN FACT FILE: HABITATS

20 FUN FACTS ABOUT TUNDRA HABITATS

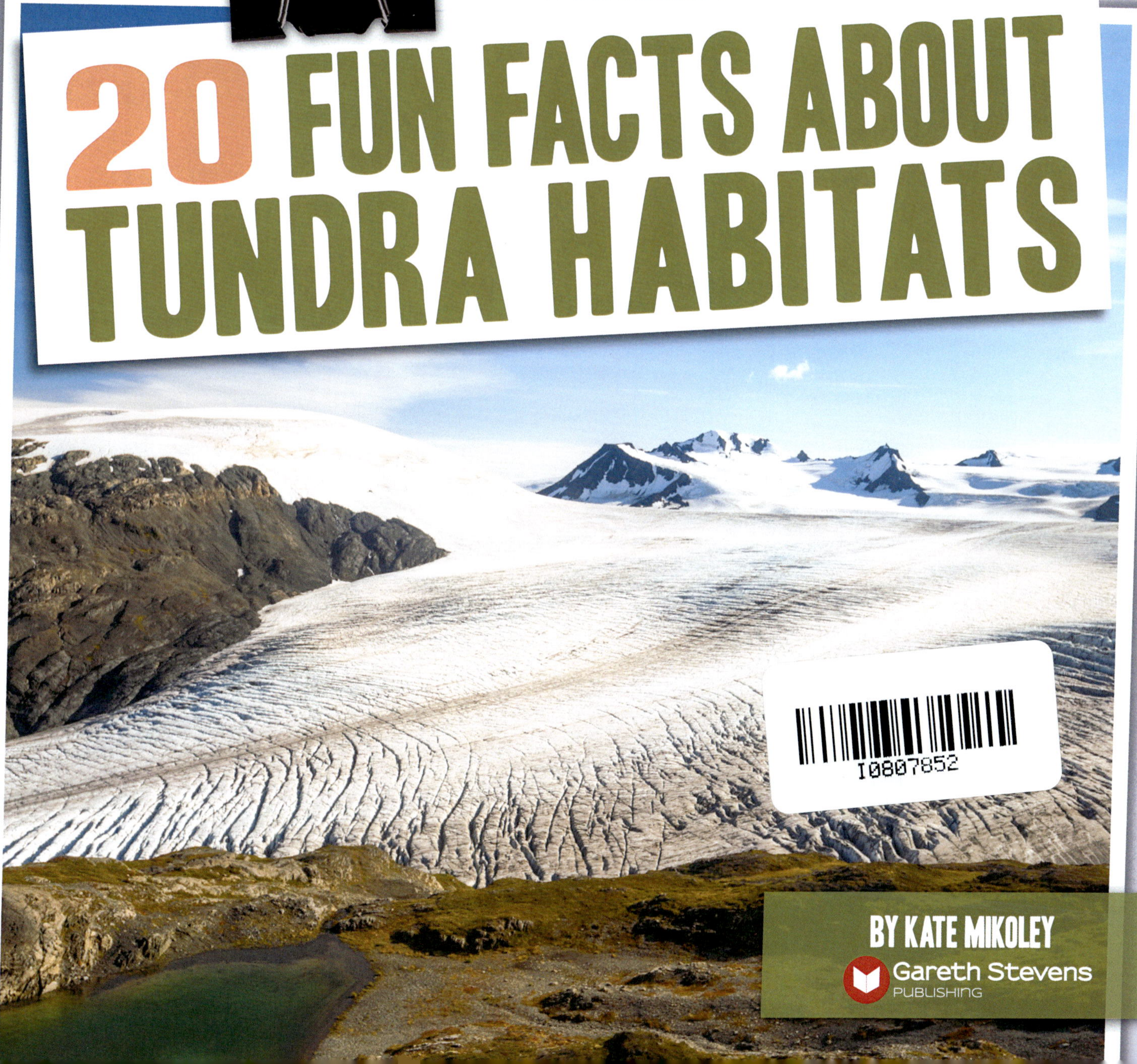

BY KATE MIKOLEY

Gareth Stevens PUBLISHING

Please visit our website, www.garethstevens.com. For a free color catalog of all our high-quality books, call toll free 1-800-542-2595 or fax 1-877-542-2596.

Library of Congress Cataloging-in-Publication Data

Names: Mikoley, Kate, author.
Title: 20 fun facts about tundra habitats / Kate Mikoley.
Other titles: Twenty fun facts about tundra habitats
Description: New York : Gareth Stevens Publishing, [2022] | Series: Fun fact file: habitats | Includes index.
Identifiers: LCCN 2020040913 | ISBN 9781538264553 (library binding) | ISBN 9781538264539 (paperback) | ISBN 9781538264546 (6 pack) | ISBN 9781538264560 (ebook)
Subjects: LCSH: Tundra ecology–Juvenile literature. | Tundras–Juvenile literature.
Classification: LCC QH541.5.T8 M55 2022 | DDC 577.5/86–dc23
LC record available at https://lccn.loc.gov/2020040913

First Edition

Published in 2022 by
Gareth Stevens Publishing
29 East 21st Street
New York, NY 10010

Designer: Michael Flynn
Editor: Kate Mikoley

Photo credits: Cover, p. 1 (main) Michael Jones/Design Pics/Getty Images; file folder used throughout David Smart/Shutterstock.com; binder clip used throughout luckyraccoon/Shutterstock.com; wood grain background used throughout ARENA Creative/Shutterstock.com; p. 5 ajlikala/Moment/Getty Images; p. 6 troutnut/iStock/Getty Images; p. 7 PeterHermesFurian/iStock/Getty Images; p. 8 SeanXu/iStock/Getty Images; p. 9 Yann Arthus-Bertrand/Getty Images; p. 10 Danita Delimont/Gallo Images/Getty Images; p. 11 SeppFriedhuber/E+/Getty Images; p. 12 Wildnerdpix/iStock/Getty Images; p. 13 Nordroden/iStock/Getty Images; p. 14 Werner Van Steen/The Image Bank/Getty Images; p. 15 Arterra/Universal Images Group/Getty Images; p. 16 Paul Souders/Stone/Getty Images; p. 17 Philippe Henry/Oxford Scientific/Getty Images; p. 18 Sophia Granchinho/Shutterstock.com; p. 19 Westend61/Getty Images; p. 20 Richard Packwood/Oxford Scientific/Getty Images; p. 21 Declan Troy/500Px Plus/Getty Images; p. 22 Kevin Schafer/The Image Bank/Getty Images; p. 23 (Arctic wolf) Adria Photography/Moment Open/Getty Images; p. 23 (caribou) Dmitry_Chulov/iStock/Getty Images; p. 23 (lichen) mgfoto/iStock/Getty Images; p. 24 Monty Rakusen/Cultura/Getty Images; p. 25 lyash01/iStock/Getty Images; p. 26 Shelley Wales/iStock/Getty Images; p. 27 Kuzhilev/iStock/Getty Images; p. 29 DEA/M. SANTINI/De Agostini/Getty Images.

Printed in the United States of America

CPSIA compliance information: Batch #CSGS22: For further information contact Gareth Stevens, New York, New York at 1-800-542-2595.

CONTENTS

Words in the glossary appear in **bold** type the first time they are used in the text.

LIVELY BIOMES

You've probably noticed that different plants and animals live in different places. For example, depending on where you live, you might have animals such as squirrels or foxes in your backyard. But other animals, such as polar bears, might not survive there.

A natural area where certain kinds of plants and animals live is known as a biome. Forests, deserts, and grasslands are a few kinds of biomes. If polar bears did live near you, chances are you'd be living in the tundra biome!

Tundra **habitats** are commonly found in cold, open areas. Don't be fooled—these places may look empty, but they're actually full of life!

WHERE ON EARTH?

FUN FACT: 1

MANY TUNDRA HABITATS ARE NEAR THE NORTH POLE.

The North Pole is the most northern point on Earth. The Arctic Circle is an imaginary line surrounding Earth near the North Pole. Tundra habitats found around or above the Arctic Circle are known as the Arctic tundra.

The Arctic tundra includes parts of the United States, Canada, Russia, and several other countries.

THE ARCTIC CIRCLE

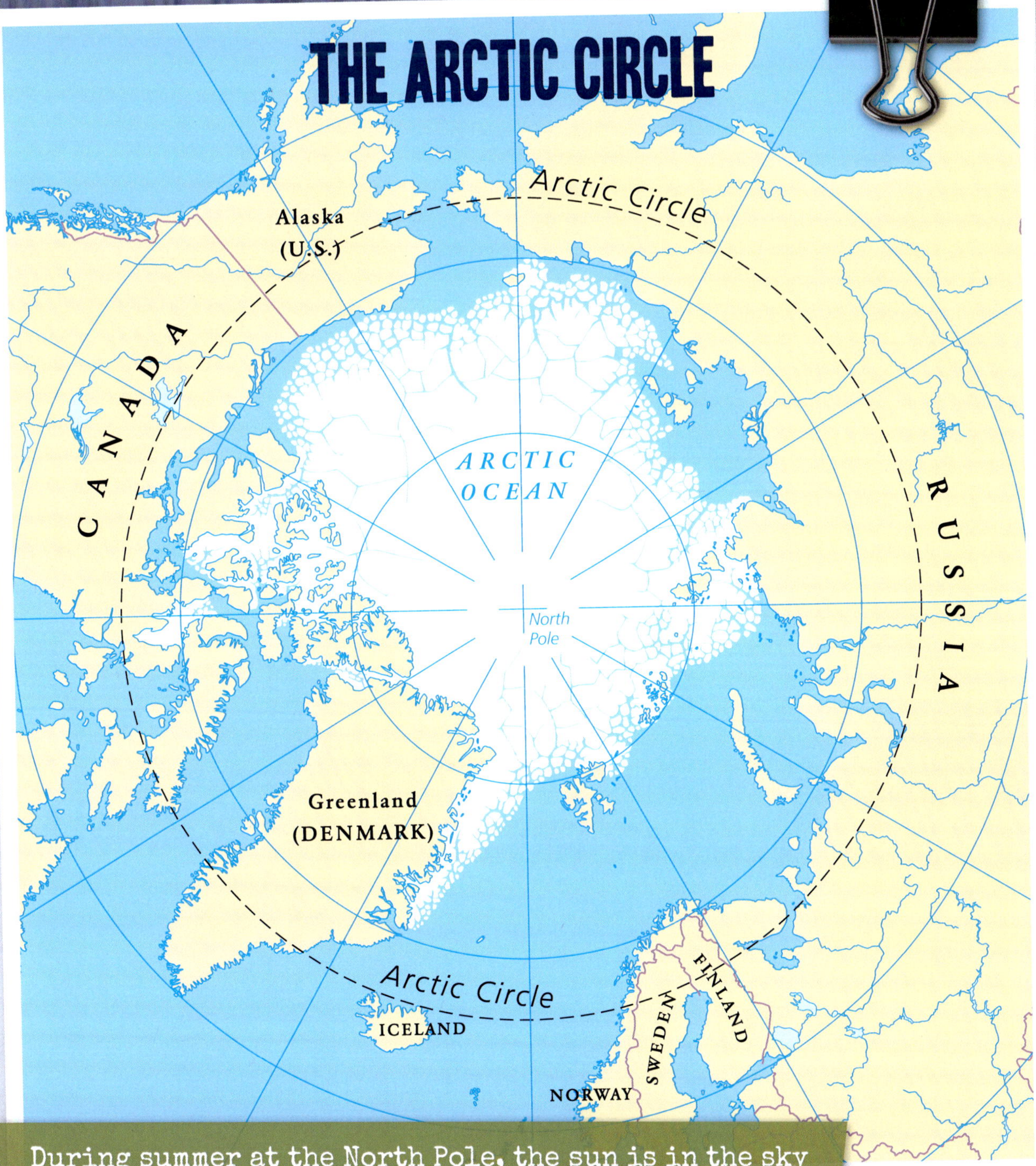

During summer at the North Pole, the sun is in the sky for 24 hours a day! However, it's usually low in the sky, so there's not a ton of sunlight.

Alpine tundra areas make up about 3 percent of Earth's land. These habitats are found on mountains all over the world.

FUN FACT: 2

SOME TUNDRA AREAS ARE FOUND AT THE TOP OF MOUNTAINS!

Trees won't grow past a certain point on very tall mountains. This is called the timberline. Alpine tundra areas are found above the timberline in places at high **altitudes**.

FUN FACT: 3

THE TUNDRA MAKES UP ABOUT ONE-TENTH OF EARTH'S SURFACE!

Some areas in Antarctica are similar to the Arctic tundra, but most are covered in ice all year. This means plants that grow in the real tundra generally can't survive here.

You might hear about the Antarctic tundra. However, many scientists don't consider this a true tundra because it can't support the kinds of life generally found in alpine or Arctic tundra habitats.

WHAT'S IT LIKE?

FUN FACT: 4

THE TUNDRA IS THE COLDEST BIOME ON THE PLANET!

Summer in the tundra can be as short as six weeks! Even then, it's rarely hotter than 50°F (10°C). In winter, the Arctic tundra can be as cold as -50°F (-45.5°C).

The alpine tundra doesn't get quite as cold as the Arctic, but temperatures still sometimes get down to 0°F (-18°C).

In summer, the top layer of permafrost may leave the tundra looking wet. It can even create small lakes. Deeper down, the ground is still frozen solid!

FUN FACT: 5

IN THE ARCTIC TUNDRA, THE GROUND IS FROZEN ALL YEAR!

Even in the summer, when the top part of the soil may **thaw**, there's a frozen layer of ground below it called permafrost. Some permafrost is 1,500 feet (457.2 m) below the surface!

At cold temperatures, water takes a while to **evaporate**. Permafrost doesn't allow much water to flow out. So even without much rain or snow, the tundra stays wet.

FUN FACT: 6

THE TUNDRA GETS LESS PRECIPITATION THAN SOME DESERTS!

There's commonly less than 10 inches (25.4 cm) of rain or snow a year in the tundra. Some deserts get as much as 24 inches (60.1 cm) per year.

FUN FACT: 7

THE WORD "TUNDRA" MEANS "TREELESS PLAIN."

Woody plants that are smaller than trees and grow low to the ground are called shrubs. Some do grow in the tundra.

You won't find trees here! Trees need deep roots. In the Arctic tundra, roots can't grow past the permafrost. Even without permafrost in the alpine tundra, it's too cold and windy for trees to survive.

POWERFUL PLANTS

FUN FACT: 10

TUNDRA PLANTS USE TEAMWORK!

Think about being outside on a cold, windy day. If you stood close together with some friends, you'd be warmer, and the wind wouldn't feel as uncomfortable. It's the same for plants. Growing close together helps them survive the tundra's **harsh** conditions.

Strong winds blow small bits of snow and ice around the tundra. Staying close together helps keep plants from being hurt by these pieces.

Tundra plants have **adapted** to survive at low temperatures. To stay warm and get light, the Arctic poppy's flowers move in the direction of the sun.

FUN FACT: 9

PLANTS HAVE TO BE SPEEDY TO SURVIVE THE ARCTIC TUNDRA.

The top layer of soil is generally free from permafrost for fewer than 90 days. Plants must grow and flower quickly before the ground freezes again.

Some flower buds stay warm by growing underground or surrounded by dead leaves. By summer, the tundra comes to life as these buds bloom.

FUN FACT: 10

SOME PLANTS CAN GROW UNDER SNOW.

Most tundra plants live several years, through multiple growing seasons. Their roots store **nutrients**, even within frozen soil. Once the top layer thaws, the plant can grow again—even if it's still under some snow!

FUN FACT: 11

LICHENS GROW ON ROCKS!

Lichens are a favorite meal for some tundra animals, such as reindeer (also called caribou).

A lichen is a plantlike organism, or living thing, made of two other organisms. A **fungus** takes in water, while an **alga** or a type of bacteria produces food. With this teamwork, lichens can survive harsh conditions. Plants called mosses grow on rocks too!

ANIMALS OF THE TUNDRA

FUN FACT: 12

SOME ANIMALS HAVE FUR THAT CHANGES TO BLEND IN WITH THEIR SURROUNDINGS!

An Arctic fox is usually a brownish-gray color in the summer. In the winter, its fur becomes white to match the snow. This helps the fox hide from predators!

Other tundra animals, including hares and some kinds of birds, change color in the winter too. Look how well this Arctic hare blends in with the snow!

Large tails cause animals to lose heat too. Lemmings are small and look sort of like mice. They have smaller ears and tails than similar animals in warmer habitats.

FUN FACT: 13

TUNDRA ANIMALS TEND TO HAVE SMALLER EARS THAN THOSE THAT LIVE IN WARMER HABITATS.

Animals can lose heat through their ears. The smaller the ears, the less heat they'll lose. Arctic hares and foxes have smaller ears than desert hares and foxes.

Many birds, such as the snow bunting, come to the tundra in the summer to lay eggs and have babies. In winter, they often move to warmer areas along coasts.

FUN FACT: 14

MOST BIRDS IN THE TUNDRA ONLY SPEND PART OF THE YEAR THERE.

During summer, many birds live in the tundra. They fly to warmer locations for winter. To prepare for the long journey, some birds eat so much that they nearly double their weight!

FUN FACT: 15

ONE KIND OF BIRD HAS BUILT-IN SNOWSHOES!

The willow ptarmigan has several layers of feathers to help it stay warm all winter!

Unlike other birds, the ptarmigan stays in the tundra all year. Feathers on the top and bottom of its feet keep it warm against snow and ice. This also stops it from sinking into the snow!

Under their thick fur, polar bears have black skin that takes in warmth from the sun.

FUN FACT: 16

MOST POLAR BEARS ARE ONE OF A SET OF TWINS.

Polar bear mothers usually give birth to two cubs at a time. The cubs spend a little over two years living with their mother and learning how to survive in the tundra.

ARCTIC TUNDRA FOOD CHAIN

PREDATORS

POLAR BEARS

ARCTIC WOLVES

SNOWY OWLS

BROWN BEARS

ARCTIC FOXES

HERBIVORES (PLANT EATERS)

LEMMINGS

CARIBOU

ARCTIC HARES

PLANTS AND PLANTLIKE ORGANISMS

FLOWERS

LICHENS

MOSSES

These are just some organisms in the Arctic tundra food chain. In general, herbivores eat plants, and predators eat herbivores. Sometimes, predators even eat other predators.

PEOPLE AND THE TUNDRA

FUN FACT: 17

TUNDRA GROUND IS FULL OF NATURAL RESOURCES.

Natural resources are things found in nature that can be used by people. Coal and oil can be found deep in the ground of the tundra. They can create power, but they also pollute the **environment**.

Oil and coal are fossil fuels, or kinds of matter that are formed over millions of years from plant and animal remains and often burned for power.

Many people worry that harm to permafrost and pollution from fossil fuels are hurting plants and animal life in the tundra.

FUN FACT: 18

HUMAN ACTIVITY CAN DESTROY PERMAFROST!

People have lived in the tundra for thousands of years. However, more recently, as oil and coal operations grew bigger, more people moved there. They put up homes, buildings, and roads, ruining much permafrost along the way.

Today, some parts of the Alaskan pipeline are raised so caribou and other animals can walk under it.

FUN FACT: 19

AN OIL PIPELINE IN ALASKA BLOCKED A CARIBOU MIGRATION PATH.

Each year, herds of caribou migrate across the tundra. When humans built a pipeline to move oil across Alaska, some caribou were unable to make their usual journey.

FUN FACT: 20

TIRE TRACKS CAN BE SEEN ON THE TUNDRA'S GROUND MORE THAN 10 YEARS AFTER THEY WERE MADE.

When soil is harmed, the few plants that can grow in the tundra may have trouble coming back during the next growing season.

The tundra is a fragile, or easily harmed, biome. Even footprints can be seen on the ground years after someone has stepped there.

TAKE CARE OF THE TUNDRA!

The tundra is an important biome—not only to the plants and animals that live there but to all the world's organisms.

Permafrost can keep plants and animals from long ago in similar states as when they lived. Scientists use these as records to learn about the past, especially about the changing weather conditions of Earth, called climate change. Using fossil fuels is one way humans cause climate change. Switching to other methods of power could help save the tundra—and the planet!

The tundra is home to many national parks. These areas can help protect plants and animals that live there.

GLOSSARY

adapt: to change to suit conditions

alga: a plantlike living thing mostly found in water

altitude: height above sea level

environment: the natural world in which a plant or animal lives

evaporate: to change from a liquid to a gas

fungus: a living thing that is somewhat like a plant but doesn't make its own food, have leaves, or have a green color. Fungi include molds and mushrooms.

habitat: the natural place where an animal or plant lives

harsh: unpleasant or difficult to experience

migration: movement to warmer or colder places for a season

nutrient: something a living thing needs to grow and stay alive

precipitation: rain, snow, sleet, or hail

thaw: to melt or stop being frozen

FOR MORE INFORMATION

BOOKS

Krajnik, Elizabeth. *Changing Tundra Environments.* New York, NY: PowerKids Press, 2020.

Quinlan, Julia J. *What is the Tundra?* New York, NY: Britannica Educational Publishing, 2019.

Topacio, Francine. *Creatures in the Icy Tundra.* New York, NY: PowerKids Press, 2020.

WEBSITES

Tundra

www.dkfindout.com/us/animals-and-nature/habitats-and-ecosystems/tundra/
Read more about tundra habitats and the animals that live in them on this interactive page.

Tundra Biome

www.nationalgeographic.org/encyclopedia/tundra-biome/
Check out more details about the tundra here.

INDEX